Annual Student Conference: Technical Program 2023

Held at the 2023 AIChE Annual Meeting

Orlando, Florida, USA
5-10 November 2023

ISBN: 978-1-7138-9323-3

Printed from e-media with permission by:

Curran Associates, Inc.
57 Morehouse Lane
Red Hook, NY 12571

Some format issues inherent in the e-media version may also appear in this print version.

Copyright© (2023) by AIChE
All rights reserved.

Printed with permission by Curran Associates, Inc. (2024)

For permission requests, please contact AIChE
at the address below.

AIChE
120 Wall Street, FL 23
New York, NY 10005-4020

Phone: (800) 242-4363
Fax: (203) 775-5177

www.aiche.org

Additional copies of this publication are available from:

Curran Associates, Inc.
57 Morehouse Lane
Red Hook, NY 12571 USA
Phone: 845-758-0400
Fax: 845-758-2633
Email: curran@proceedings.com
Web: www.proceedings.com

TABLE OF CONTENTS

BEYOND THE BACHELOR'S: MASTERING GRADUATE SCHOOL AND FELLOWSHIP APPLICATIONS

GREEN COLLAR CAREER OPPORTUNITIES FOR CHEMICAL ENGINEERS

MIXING OF VISCOUS, NON-NEWTONIAN FLUIDS

FUND YOUR PH.D. AT THE UNIVERSITY OF KANSAS WITH THE SELF GRADUATE FELLOWSHIP

ENGINEERING VALUE - THE IMPORTANCE OF CREATING AND IDENTIFYING VALUE IN THE WORLD OF CHEMICAL ENGINEERING

CREATE YOUR UNICORN CAREER!

FINANCES AND YOUR FIRST JOB – FINANCIAL PLANNING FOR CHEME CAREERS

THE PATH TO BECOMING A LICENSED PROFESSIONAL ENGINEER

WISE INTERNSHIP: AN OPPORTUNITY TO APPLY ENGINEERING TO POLICY IN D.C.

The Ins and Outs of Planning a Student Regional Conference

Jason Hillman

Planning a regional conference is a daunting task for a student chapter to undertake, but we are here to help. The University of Florida recently held the Southern Regional Conference, and this workshop will cover exactly how to plan your regional conference. We will discuss everything that needs to be done to have a successful conference, starting from the day you win your bid to host the conference up to the execution. Along the way, we will be offering our advice and recommendations on how to make the more complex aspects of the conferences, like the ChemE Car and Jeopardy competitions, run smoothly.

What It Takes to Successfully Shape Your K-12 Event/Program

Olivia Dohm, Madeline Mueller, Eleanor Lopez, Ashley Pennington, Fernando Mérida Figueróa, J. Patrick Abulencia

In this workshop, attendees will learn how to account for logistics and coordination necessary to run an outreach event or program. During the workshop, attendees will learn by doing as they fill out the AIChE K-12 STEM Minigrant application, and any attendee who wishes to submit the grant afterward may do so or take it back to their student chapter/local chapter and learn from the process.

The presenters will speak on their own experiences of hosting AIChE STEM outreach events and what goes into starting a K-12 program of that level. A checklist will be provided to help people interested in hosting smaller scale events at their school/in their communities. Through participation in this how-to workshop, attendees will hone skills and learn the benefits (to yourself, and your community) of volunteering in outreach programs, inspiring K-12 youth to consider and pursue STEM, and growing leadership skills through volunteer organizations.

Building a Student-Led Safety and Ethics Culture: Lessons from a Student-Run Conference

Michael Reichmann, Annie Ellis

This workshop will be presented by a team of students at BYU that has organized a safety and ethics conference for the past two years. This has been a huge success, with over 300 participants, keynote speakers, workshops, panel discussions, lab tours, games, competitions, and networking with around a dozen companies present. We plan to discuss how this conference has been used to develop and strengthen a safety and ethics culture among our students, provide leadership opportunities, and enhance ties with industry. We'll present the organizational structures, fundraising, communication, and logistics of planning and execution. We'll also discuss the motivation for the conference and how this is one option of several for developing a safety and ethics culture on campus. Here's a link to the conference website: https://engineeringsafetyethics.byu.edu

Organizing a Successful Research Symposium: Highlighting Undergraduate Research

Evan Grothen, Shriya Khandelwal, Isabel Panicker

This workshop will detail the planning of a research poster symposium and the benefits it can offer to your student chapter and students. We will discuss strategies on attracting a high attendance of students and faculty and ensuring a valuable experience for all attendees. If time permits, there will also be a discussion on do's and don'ts of making a research poster. The winner of Purdue AIChE's poster symposium will also be speaking on her experience. This will be useful for student chapters looking to organize new events and current students interested or participating in research.

From Campus to Career: Chemical Engineering Pathways Unveiled

Carmen Daoud

Brief description of workshop:

This session will feature a panel of ChemE's from the Young Professionals Committee (YPC) who are currently pursuing a variety of different career paths, from graduate school to industry to government jobs. Panelists will discuss what a typical day looks like in their current roles, career decisions they made to get there, uncertainties they pushed through, and pros and cons of different careers. This panel will provide a great overview of many different career options for ChemE's. Panelists will also discuss how AIChE engagement after graduation helps in their unique career path.

A Day in the Life of a Process Engineer

Cynthia Murphy-Ortega

What does a Process Engineer do at an energy company? What are some of the daily tasks and challenges that they work on? How do they enable lower carbon operations and support the energy transition? What skills (beyond engineering) do they find most valuable? How have they transitioned from an academic to professional environment? Join Process Engineers from Chevron as they describe their roles, share what they have learned, and explain how they have integrated teamwork, communication, technology, and sustainability into their careers.

Materials Engineering and Sciences in AIChE

Rafael Verduzco, Adrianne Rosales

The MESD workshop will feature speakers across all areas of the Materials and Engineering Sciences Division, including polymers, biomaterials, inorganic materials, electronics and photonics, and composites. Representatives from each area will present representative research to give students an overview of research programs in MESD. Speakers will also present advice to students as they pursue and consider different PhD graduate programs or other career options. After the presentations, speakers will be available for a social "meet and greet" with students.

Innovations in Materials and Pharmaceutical Manufacturing

Nastassja Lewinski, Thomas Roper

According to the U.S. Food and Drug Administration, over half of American hospitals changed patient care or delayed treatment due to drug shortages. A major cause of drug shortages is attributed to manufacturing quality problems. VCU Engineering has been investing in the advanced manufacturing space for years, and as a result, has been recognized for its innovation and expertise in pharmaceutical engineering through research and education and the establishment of the Medicines for All Institute (M4ALL) and the Center for Pharmaceutical Sciences and Engineering. This workshop will discuss the innovative research being conducted to optimize materials and pharmaceutical manufacturing and opportunities to participate in these efforts to improve access to high-quality medications across the globe.

A Roadmap for Career Success: How to Set Your Course

Larry Lindsey

Investing in your career development should be one of the key priorities as you contemplate and plan your professional future. This Career Development workshop will teach you how to create a highly effective resume, polish up your LinkedIn profile, interview with confidence, and how to build a lasting professional network.

Starting Your First Plant Position in Industry

Carmen Daoud

What do employers look for in a recent college graduate? How can you prepare to ace your industry interviews? What does the typical day of an engineer look like in plant position? How can you make the most of your internship experiences to propel you into a full-time job? How can AIChE help you land a job and excel in your first few years out of college? Join the Young Professionals Committee (YPC) in a discussion on preparing to apply for plant jobs and excel in your first position. YPC members will also discuss what they wish they knew when starting their first jobs. The session will feature recent-graduate young professional engineers from various companies talking about their experiences and giving advice on making the transition from college to career.

Survive and Advance with Confidence: Navigating the Ph.D. Labyrinth

Reginald Rogers Jr.

A PhD is more than just getting three letters behind your name. It is a journey where you are learning more about yourself as you become an expert in your chosen field of study. However, no two PhD journeys are the same. From Year 1 to Year 5, there will be many crossroads one will encounter before reaching the finish line. Each crossroad may leave you asking: "Is the PhD right for me?" Such a question is common among individuals pursuing the ultimate advanced degree. In this workshop, we will walk through the PhD journey from start to finish. How do you tackle each year of your program? How you negotiate difficult advisors? What do you do when you begin to feel the "PhD Fatigue"? How do you leverage your PhD experience in developing the next phase of your career? The nuts and bolts to tackling a PhD will be the same across the board, but the end result will depend on YOU. This interactive workshop will give you the chance to think about what you need to do in order to be successful and advance to the next level...with confidence!

Learning Outcomes and Take-Aways

By the end of this workshop, participants should be able to:

1. Understand the complexities of navigating a PhD with a different mindset compared to their undergraduate degree program
2. Write out a plan of action for successfully navigating their PhD journey
3. Discuss, freely, issues they are currently facing in their PhD program and how to overcome those issues
4. Walk away confident in their efforts to complete their PhD with a high level of success

Chemical Engineers in the Pharmaceutical Industry

Emily Cunningham, Jingyao Wang, Stephen Ma

This session is designed to introduce students to the roles chemical engineers play in pharmaceutical discovery, development, and manufacturing. It will feature a brief presentation on the pharma industry followed by Q&A with the panelists. Students are encouraged to bring their questions!

The World of Particle Technology

Ben Freireich

Billions of pounds of bulk solids are processed and handled every year by the US process industries, yet most chemical engineers are ill-equipped to deal with the complexities of engineering science of solids processing. Hence, plants and products suffer with lost production, inability to achieve design production rates, off-grade or off-specification products. During this session, we will take a look at the fun and exciting world of solids processing. Specifically, we will look at some of the more common particle-based technologies examining both the important role they play in society today along with the associated technical challenges.

Imperial College London: Your future self is waiting for you

Karen M. Polizzi, Ronny Pini

Interested in taking the next step in your career in one of the world's leading Chemical Engineering Departments? Join us to find out about the postgraduate opportunities offered by the Department of Chemical Engineering at Imperial College London. Situated at the heart of London, ranked the best student city in the world, Imperial College London is a global institution with a multicultural study environment and a vibrant student community.
The event will include information on:
1. the 1-year MSc programme in Advanced Chemical Engineering and the Biotechnology, Materials Engineering and Process Systems Engineering streams;
2. the PhD programme that covers the broad range of research themes hosted by our Department;
3. the postgraduate admissions process for MSc and PhD studies;
4. the scholarship opportunities offered by our Department.
The event will feature presentations by the MSc course Coordinator, Prof Karen Polizzi, and Postgraduate Admissions Tutor, Dr Ronny Pini, followed by a Q&A session. It is aimed at current undergraduate and MSc students working towards a degree in Chemical Engineering, Materials Engineering, Chemistry, and other related disciplines who are planning to undertake further technical and research training. It will also be relevant to Engineering and Science professionals wishing to develop their technical and research skills and obtain further qualifications.

SLB New Energy: Accelerating the transition to low carbon energy.

Adel Boussadia

We are SLB, a technology company driving innovation for a balanced planet. We thrive to decarbonize the energy industry and to scale new energy systems. Come meet our team and explore opportunities to build your career.

It's Not Bragging: The Art of Self-Promotion in the Job Search and Career Plan

Alaina Levine

Self-promotion is not bragging- as long as you share truthful information about how you can help another party, at the appropriate time and place. In this session, we will discuss the necessity of self-promotion and how to do it effectively, professionally, and appropriately across cultures, communities, and sectors. Communicating and promoting your value is an essential component of advancing your career - and advancing the careers and missions of your partners. Learn how to do it the right way!

Prioritizing mental health as an engineering student

Sarah Wilson

Research on mental health in engineering students shows that engineering students: 1) feel that they don't have time to prioritize their mental health, 2) are less likely to seek help for their mental health, 3) feel that they would be more likely to seek help if they received help from a friend or peer and 4) normalize the stress of the engineering training environment (Lipson et al. 2016; Jensen et al. 2023; Jensen and Cross 2021; Wright 2021; Ban et al. 2022). Therefore, this workshop will talk to students about the importance of prioritizing mental health through the coverage of 4 key topic areas: 1) why do engineering students need to prioritize mental health?, 2) the importance of advocating for yourself and your classmates, 3) recognizing normal stress vs. mental health distress and 4) identifying mental health resources on your campus. The workshop will equip students with the skills and knowledge to prioritize their mental health and access mental health resources within their college campus.

Beyond the Bachelor's: Mastering Graduate School and Fellowship Applications

Juliana Yang

Are you an aspiring chemical engineer ready to take the next step in your academic journey? Join us for this informative session where we will dive into the intricacies of applying to graduate school and fellowships. From crafting a compelling personal statement to securing strong letters of recommendation, we will provide you with expert advice and proven strategies to maximize your chances of acceptance. You will also learn about the various funding opportunities available through fellowships and gain valuable insights on how to navigate the application process for them. Whether you're considering a Master's or Ph.D. program, this session will equip you with the tools and knowledge to confidently pursue your academic goals and launch a successful career in chemical engineering.

Green Collar Career Opportunities for Chemical Engineers

Yinlun Huang, Kerry Kelly, Gerardo Ruiz-Mercado, Richard Siegel, Tony Wu

This workshop will engage students in an interactive discussion about challenges, advantages, and opportunities for chemical engineers in the environmental field. A brief overview presentation of opportunities in industry, academia, consulting, and government will be followed by an extensive question and answer session to help the students better understand their potential career options.

Mixing of Viscous, Non-Newtonian Fluids

Richard Grenville

Many viscous fluids processed in industry exhibit non-Newtonian rheology. These industries include paints, personal products such as shampoo and detergents and food products such as ketchup and mayonnaise. In some cases, the end use properties of the products are highly dependent on their rheologies. For example, the sensation of eating some foods is partly a result of its texture or "mouth feel" and the of performance of coatings will depend on how they flow when they are applied to a surface.

In this seminar the measurement of viscosity will be discussed with the interpretation of the data. Then the application of this information to the design of mixing equipment, where necessary, will be presented.

The power drawn by an impeller operating in the laminar regime is proportional to the viscosity of the fluid (as is pressure drop in laminar pipe flow) but, if the fluid is non-Newtonian, its viscosity will be determined by the shear rate that the impeller imparts to it and this must be taken into account in the sizing to the agitator drive.

Another process where this behavior must be taken into account is the blending of non-Newtonian fluids. The apparent viscosity of the fluid will be low near the impeller, where velocity gradients are high, and high near the vessel wall where velocity gradients are low and the difference in viscosity in these regions must be accounted for in order to make an accurate estimate of the blend time.

Finally, there is a class of fluids which exhibit a yield stress, meaning that a minimum level of shear must be applied before the fluid will move. Examples include toothpaste and high solids loading slurries. The rules used to size mixing equipment for these fluids will also be discussed.

Fund your Ph.D. at The University of Kansas with the Self Graduate Fellowship

Daniel A. Rivera

Embark on an extraordinary academic journey with the Self Graduate Fellowship at the University of Kansas. This prestigious program offers unparalleled resources, mentorship from renowned faculty, and a dynamic community of fellows. Gain invaluable research experience, access cutting-edge facilities, and pave the way for a successful career. Unleash your potential at KU's Self Graduate Fellowship!

Engineering Value - The Importance of Creating & Identifying Value in the World of Chemical Engineering

Ryan Morrison

Are you ready to take your chemical engineering skills to the next level? The ability to identify and understand the value of various chemical engineering tasks is such an important skill in industry. Being able to properly justify the benefits of a capital project, process optimization study, or incident investigation can lead to much higher rates of success. Join us for a seminar on understanding various types of value creation and how to determine the qualitative and/or quantitative value to a company in your chemical engineering work.

Create Your Unicorn Career!

Alaina G. Levine

In this session, we will discuss how to leverage your talents and loves, as well as networking to create a unique, customized career, which we call Your Unicorn Career. A Unicorn Career honors your authenticity and allows you to solve the problems that bring you meaning, use the skills that bring you joy, and get paid! We will address how to understand your unique value, skills and interests and how to align that with employers who will hire you. We will discuss a data-driven and process engineering approach to create opportunities for Your Unicorn Career to come to fruition.

Finances and Your First Job – Financial Planning for ChemE Careers

Carmen Daoud

How do I start saving money? What benefits should I enroll in? How do I start planning for retirement? Join the Young Professionals Committee (YPC) for a discussion on the benefits and rewards of financial planning when starting your first job. This session will feature YPC members in a variety of careers (graduate school, industry, government, etc.) and discuss tips for money management and planning for retirement. YPC members will discuss topics such as investing in 401Ks, budgeting, benefits packages, student debt, and planning for retirement. YPC members will also discuss topics that they wish they knew when starting their first jobs.

The Path to Becoming a Licensed Professional Engineer

Joseph Cramer, William Parrish, John Wagner

This presentation discusses how becoming a Professional Engineer (PE) can enhance a chemical engineer's career. Students learn about the PE licensure process and the two exams required to become licensed. The Computer Based Testing (CBT) exam content for both the Fundamentals of Engineering (FE) Exam and Practice of Engineering (PE) exams are described. The presentation utilizes embedded videos of young PEs and stresses the value of taking the FE exam before, or shortly after, graduation. It aims to demystify the FE/PE process. The workshop leaders are Joe Cramer, Bill Parrish and John Wagner who are chemical engineering PE's and active members of the AIChE's Licensing and Professional Development Committee as well as the National Council of Examiners for Engineering and Surveying's Chemical PE Exam Committee or FE Exam Committee. Joe and Bill have also served on AIChE's Career and Educational Operating Council which oversees licensing for AIChE. John has been active for many years on the NCEES FE Exam Committee. The workshop encourages Q&A and interaction with attendees.

WISE Internship: An Opportunity to Apply Engineering to Policy in D.C.

Dr. Martin Abraham, Courtney Cochran, Aicha Sama, Heather Yuengling, Evan Erickson

The Washington Internships for Students of Engineering (WISE) program is a collaboration of several professional engineering societies who select students for a summer internship in Washington D.C. The interns research and write a public policy paper on a topical engineering-related issue and present them on Capitol Hill. The interns have remarked on the invaluable experience of learning how to develop a public policy paper, lobby for legislation, and develop a network of contacts that often lead to future endeavors.

Dr. Martin Abraham, Past Chair of the PAIC and a WISE Mentor, will introduce Aicha Sama, Brown University, Courtney Cochran, Mississippi State Univ., and Evan Erickson, Univ. of Wisconsin-Madison, who were the 2023 AIChE WISE Summer interns. The Interns will discuss their experiences as a WISE intern and present a short synopsis of their papers.

Navigating Dynamic Careers in Catalysis and Reaction Engineering

Jesse Bond

Panelists will discuss their careers in the fields of catalysis and/or reaction engineering; how the Catalysis and Reaction Engineering Division of AICHE has played a role in their careers; and advice on how to navigate a dynamic and fulfilling career in the field.

Building a Sustainable Future in Chemical Engineering

Mihaela Hahne, Bruna Giron

In today's dynamic market, the demand for sustainable processes is ever-increasing, pushing the boundaries of traditional engineering. Aveva Process Simulation presents an innovative, integrated platform that empowers industries to design sustainable processes at the speed the market demands, all while paving the way towards a circular and eco-conscious world.

Join this workshop to learn more about AVEVA's innovative solutions in process simulation and XR technology that are driving sustainability and operational efficiency simultaneously. These advanced tools empower industries to meet the ever-evolving demands of the market while actively contributing to the global pursuit of sustainable and environmentally responsible processes.

Networking for Nerds: Find and Build Win-Win Relationships That Rock

Alaina G. Levine

Networking is the most honorable and valuable endeavor in which you can engage, because it is built on a spirit of generosity. At its core, networking is all about crafting win-win alliances where both parties provide value. In this session, we will discuss strategies and tactics for finding new collaborators and building mutually-beneficial partnerships with professionals at in-person meetings and across the globe. We will practice our networking and you will discover actual wording and questions to use to spark meaningful conversations. We will also explore networking using digital platforms and how to engage with people you meet at virtual events. You will emerge more confident and comfortable in your networking- now that you know exactly how to do it. Let's network!

I Heart Failure: Recover and Discover from Professional Loss

Alaina G. Levine

Everyone fails multiple times in their lives. Perhaps you didn't get that job or get named to a choice project, or the project you did lead just did not achieve its expected goals. Maybe your team didn't rally as you envisioned and became dysfunctional. Whatever the situation, there are ways to manage and achieve success from failure, as long as you plan ahead for setbacks. And even if you experience challenges in a way you did not anticipate, there are still other ways to turn things around. Failure can be absolutely fabulous in our career paths and organizations and how to create mechanisms so that we can spot, address, learn from, and move on from these setbacks. We don't have to welcome failure with open arms, but we can certainly learn to appreciate the gift that failure gives us by understanding the lesson to make us stronger and more agile, creative, communicative, and ultimately more victorious in our careers, teams, and organizations.

LGBTQ+ & Allies Panel: Navigating Your Next Steps as an LGBTQ+ Undergraduate

Bryan Deschamps

Undergraduates have a lot of questions for what comes next in their careers. Whether it be graduate school, industry, or another path, there is a lot to learn and navigate. These anxieties can be multiplied for LGBTQ+ undergraduates. This panel session will have recruiters and young graduates giving advice and best practices for navigating recruiting and getting started for whatever comes next.

How to Make the Most of Your Omega Chi Epsilon Chapter

Christi Luks

Officers and members of Omega Chi Epsilon, the National Chemical Engineering Honor Society, are encouraged to attend this session to meet with national officers and share best practices and favorite activities. Students from schools that do not have an Omega Chi Epsilon chapter are also invited to attend to learn more about establishing a chapter on their campus.

AUTHOR INDEX